AF710604

E1–GEOMETRY

SCRATCH YOUR BRAIN WHERE IT ITCHES

MATH GAMES, TRICKS, AND QUICK ACTIVITIES

DOUG BRUMBAUGH

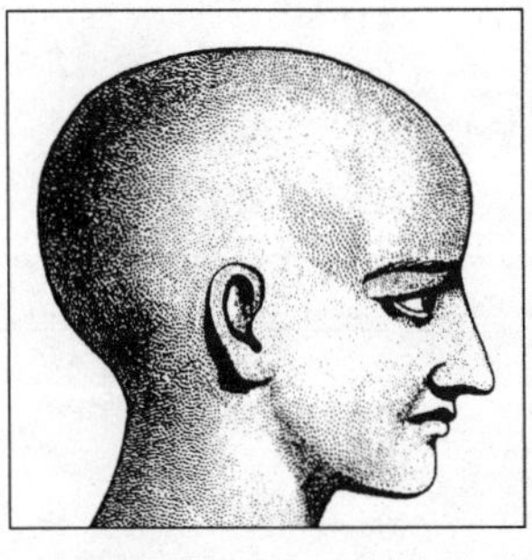

CRITICAL THINKING BOOKS & SOFTWARE
www.criticalthinking.com
P.O. Box 448 • Pacific Grove • CA 93950-0448
Phone 800-458-4849 • FAX 831-393-3277
ISBN 0-89455-526-X
Printed in the United States of America

Contents

PICKY PUZZLERS

1. Remove two toothpicks from the given arrangement so that the new arrangement will consist of only four congruent squares. Each side length of any square must equal the length of the whole toothpick.

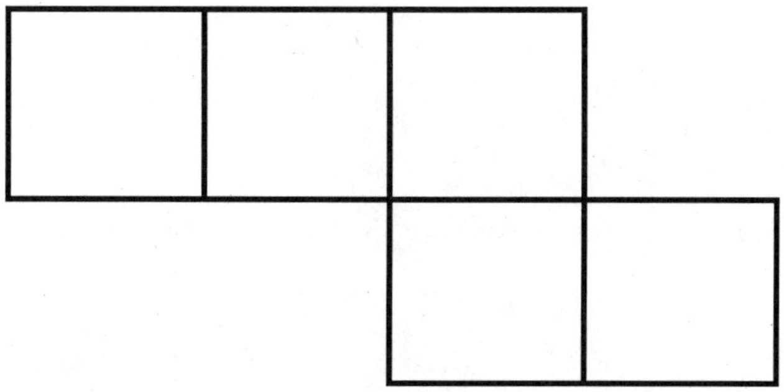

2. Move two toothpicks from the given arrangement so that the new arrangement will consist of only four congruent squares. Each side length of any square must equal the length of the whole toothpick. All toothpicks must be used.

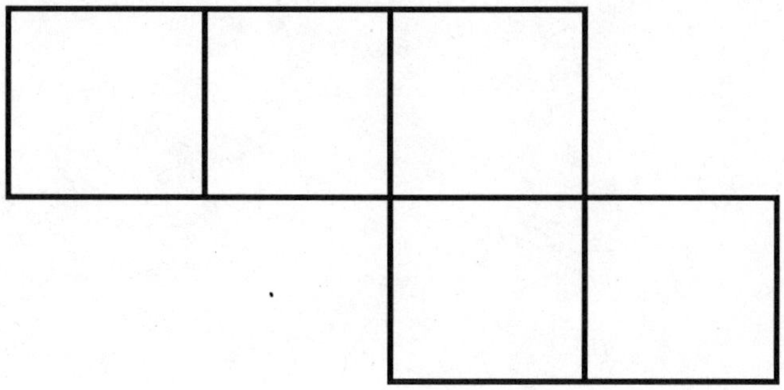

3. Use 9 toothpicks to form 3 squares, each side of which is a complete toothpick length.

4. Make a figure that consists of two squares and four triangles from 8 toothpicks without breaking or bending them.

5. Using 24 toothpicks, how many squares can be made using six toothpick lengths on a side? Five toothpick lengths on a side? Four toothpick lengths on a side? Three toothpick lengths on a side? Two toothpick lengths on a side? One toothpick length on a side? For some of these, you may not be able to use all the toothpicks.

6. Arrange six toothpicks to form 4 congruent equilateral triangles. The side length of each triangle must equal the length of a whole toothpick.

7. Thirteen toothpicks are used to create a large rectangle composed of 6 small congruent rectangles. Using twelve toothpicks, create another shape that is composed of 6 small congruent shapes.

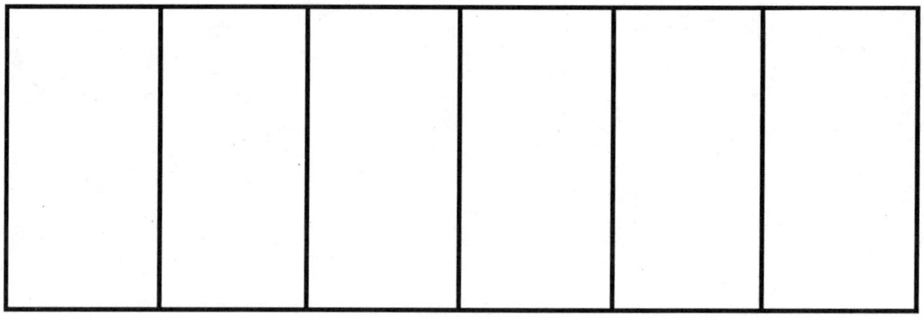

8. A rhombus is formed by 2 congruent equilateral triangles sharing a common side. The midpoints of each of the sides of the equilateral triangles are connected, creating 8 smaller congruent equilateral triangles. Use toothpicks to form each side of the small triangles then remove 4 toothpicks from this figure and leave exactly 4 triangles.

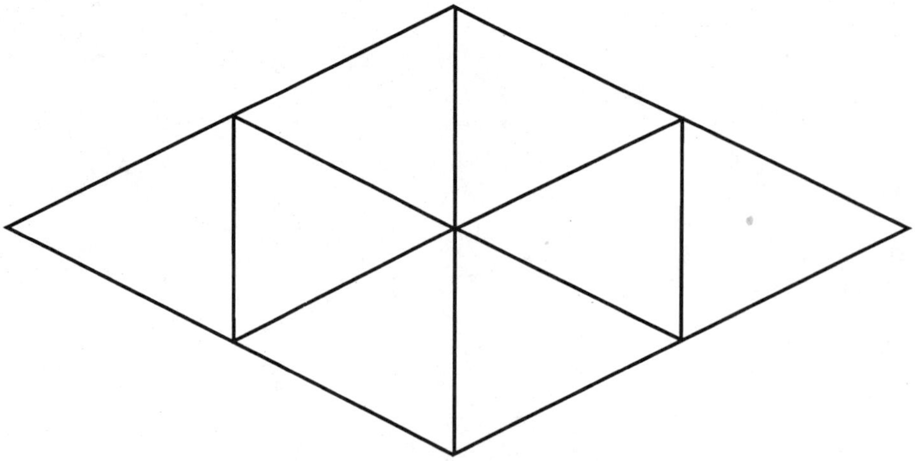

Plane Figures

9. Describe an in-line skate. In your description, use appropriate geometry terminology.

10. Make a tool like the one shown. The line segments are strings. The piece with the circular hole in it is made from cardboard. Make similar pieces with different holes: square, equilateral triangle, isosceles trapezoid, parallelogram. Predict what the projection of each hole will be. Use the tool to check your prediction. Explain your result.

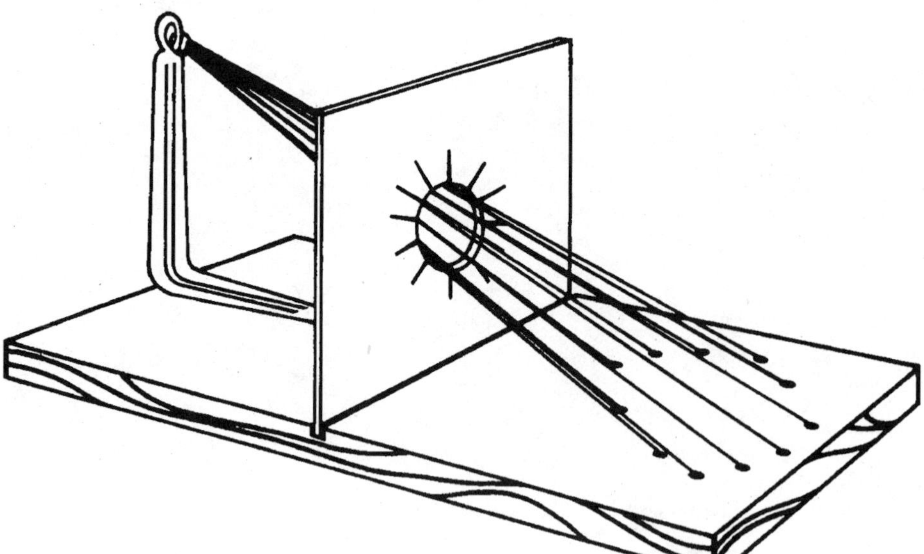

11. How many squares are on a checkerboard?

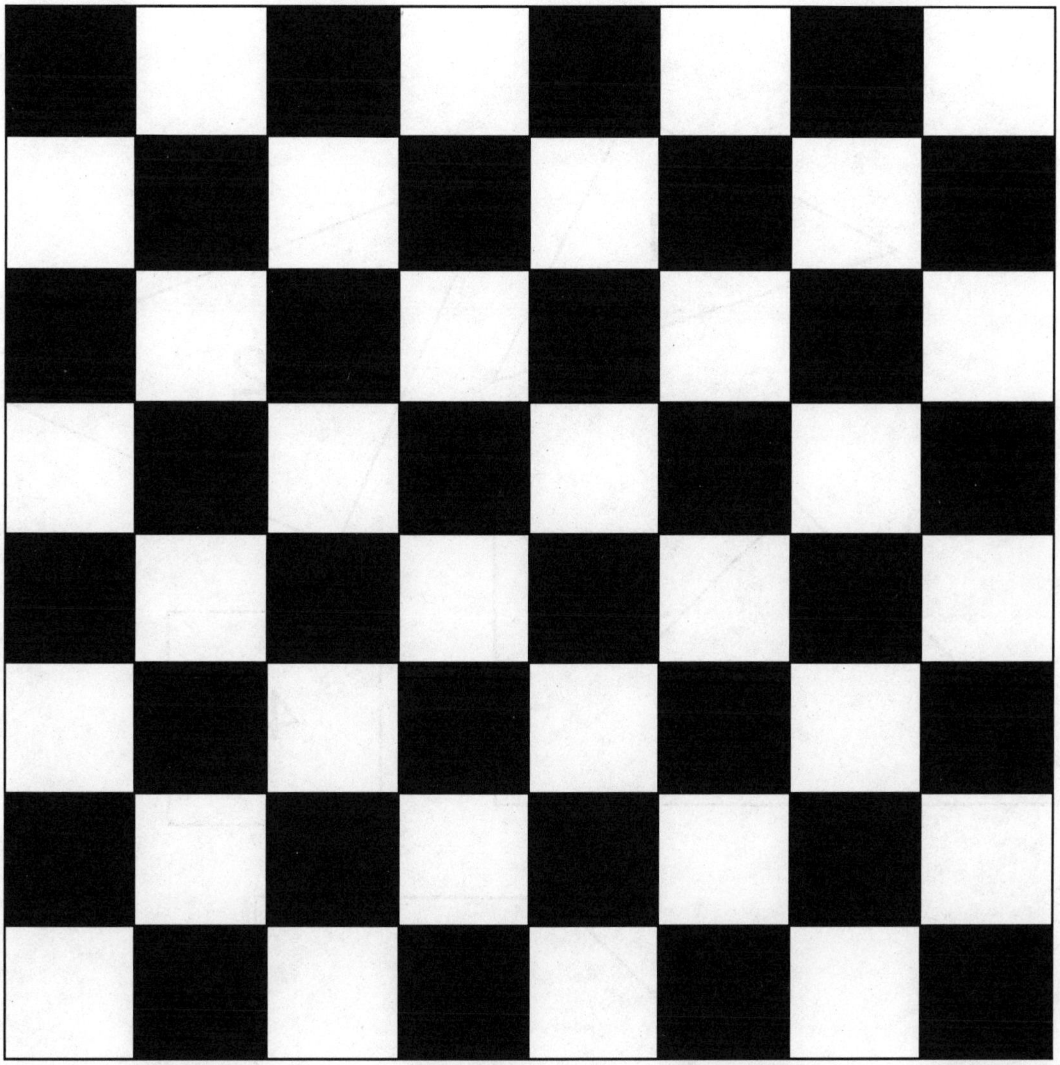

12. These 5 pieces can be put together to form a square. They are drawn to scale. You may find it helpful to cut the pieces out of cardboard or poster board.

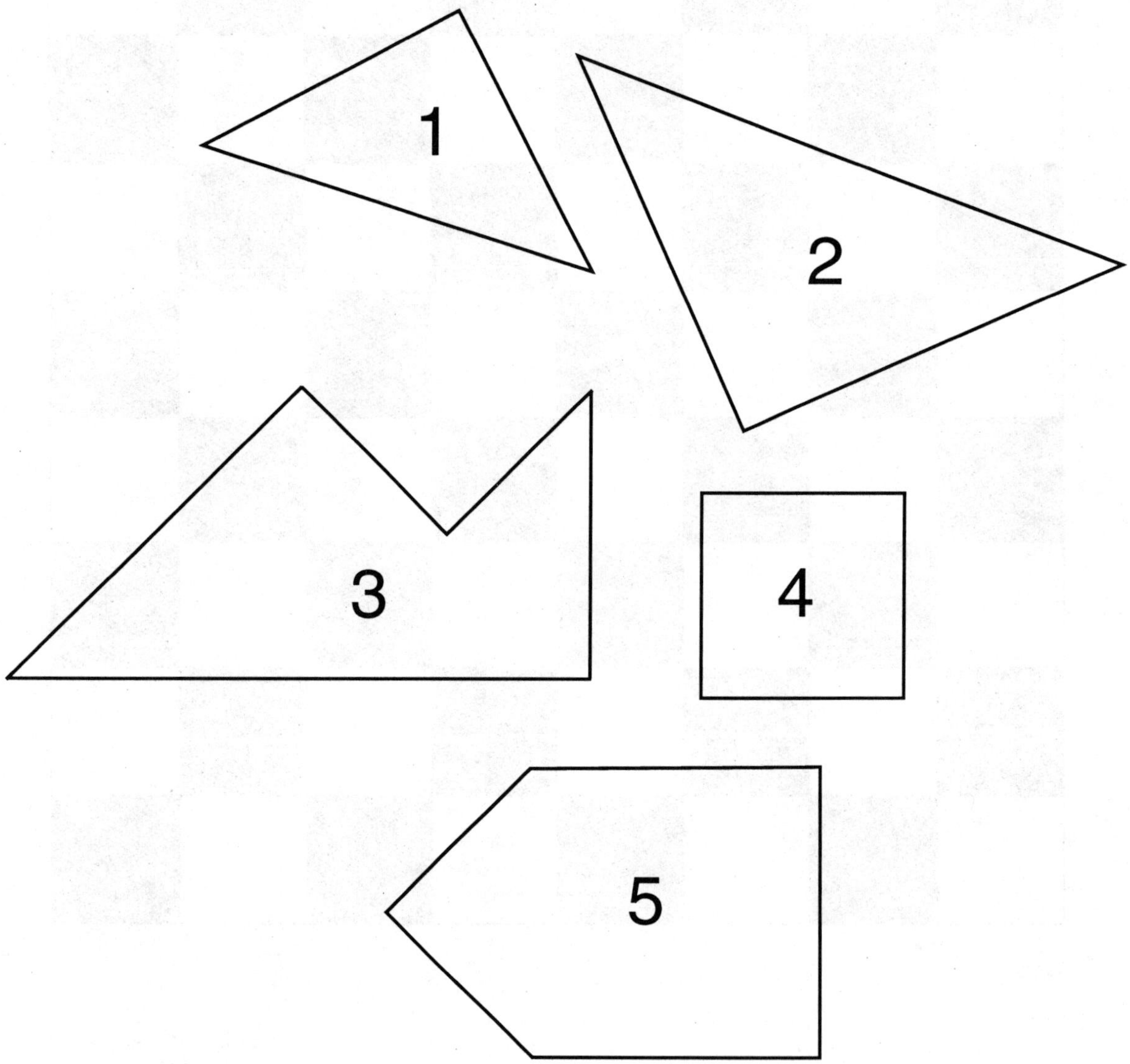

13. How can a gardener plant ten trees in five rows, having four trees in each row?

14. A farmer wants to buy a plot of ground adjacent to property currently owned. A real estate agent says the triangular plot has side lengths 30, 45, and 85 decameters. The land costs $17.99 per square decameter. Knowing that Heron's formula can be used to compute the area of any triangle, how much did the farmer pay for the ground?

Heron's formula:

Area = $\sqrt{s(s-a)(s-b)(s-c)}$ where a, b, and c are the triangle side lengths and $s = \frac{(a+b+c)}{2}$

15. Arrange four points on a plane so the distance between any two points is one of two possible lengths; for example, one possible solution would be to have the four points be the vertices of a square. The side lengths of the square would be one length. The diagonals would be the second length. There are 5 other possible solutions to this question. How many can you find?

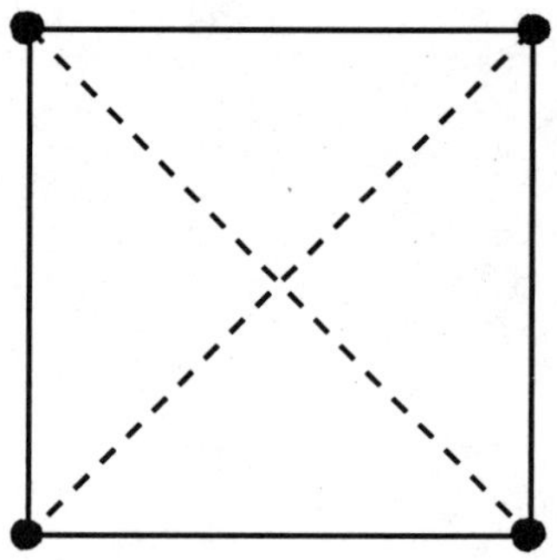

16. We had a window that was a yard square. We thought it would let in too much light, so we framed in half of it. When we got done, we still had a window that was a yard high and a yard wide. How did we do it?

17. Each of the 6 figures below is made up of 5 congruent squares. All 30 of the squares are congruent. What questions could you ask about these figures? What are the answers?

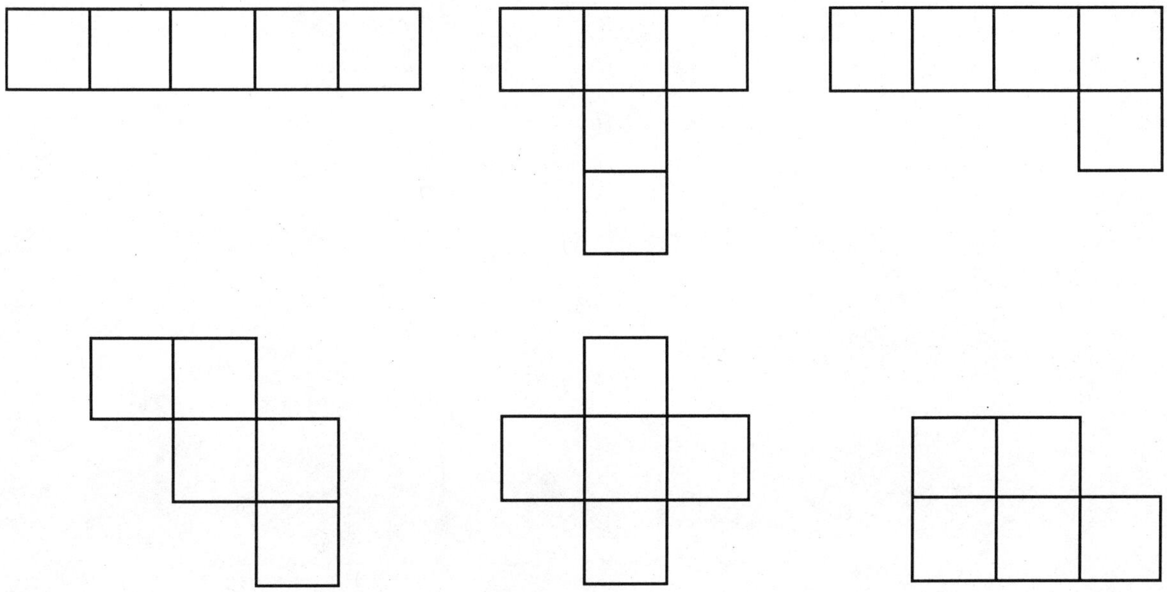

18. Calendars consisting of 4 cubes are commercially available, but they are also relatively easy to make. One cube serves to represent all months. One cube serves to represent the days of the week. The final 2 cubes are used to show the dates of the month. Design the cubes of this calendar.

19. A 9-by-12 rectangular rug has a 1-by-8 hole in its center, the hole being 4 feet from either long side and 2 feet from either short side. The sides of the hole are parallel to the sides of the rug. Cut the rug into exactly 2 pieces which can be sewn together to make one 10-by-10 square rug.

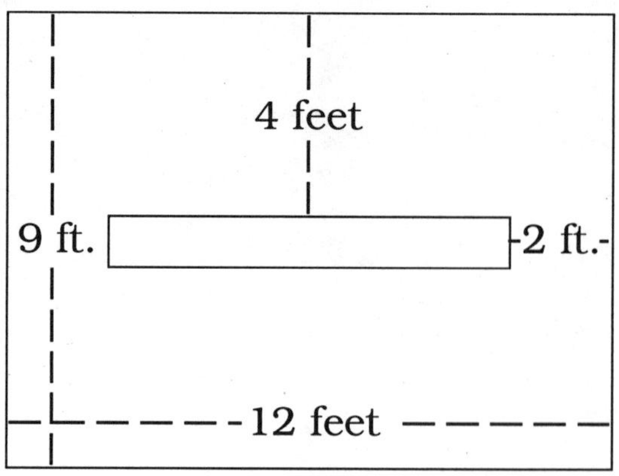

20. Why are manhole covers round?

EXTENSION: Is there any other shape that will work? Assume the cover and the hole are the same shape, the hole is at least a meter across at the narrowest point, and if the hole and lid were concentric figures, the lid would be slightly larger (1.02 meters across at the narrowest point) than the hole.

21. Gather several cylinders and a tape measure or calipers. A calculator will be useful for this activity. Measure the circumference and diameter of each of the cylinders. Divide the circumference by the diameter. Arrange the results in a table and write a conjecture to accompany the information shown in the table.

ITEM	CIRCUMFERENCE	DIAMETER	CIRCUMFERENCE ÷ DIAMETER

22. Cut a circle into an even number (8 or more) of congruent pie-shaped pieces with the length of each straight side being equal to the length of the radius of the circle. Put the pieces together as shown in the picture. Notice that the shape looks like a parallelogram. Imagine that the edges made up of curves are straight (the more pieces used, the straighter the long side appears). What is the length of each side of the "parallelogram"? What is its height? What is its area? Since the "parallelogram" was made from a circle, what is the area of the circle?

23. A kid is full after eating a round six-inch-diameter pizza. Assuming there are two kids with the same eating capacity, what size pizza (only one) should they order so that they will both be full and there will be no leftovers?

24. A person selling pencils packaged them in bundles. A bundle measuring one inch in diameter sells for a dollar. At the same rate, how much would a two-inch bundle sell for?

25. A solid is viewed as shown. This view is the same for both the top and front. Draw the side view. How many different solutions can you generate?

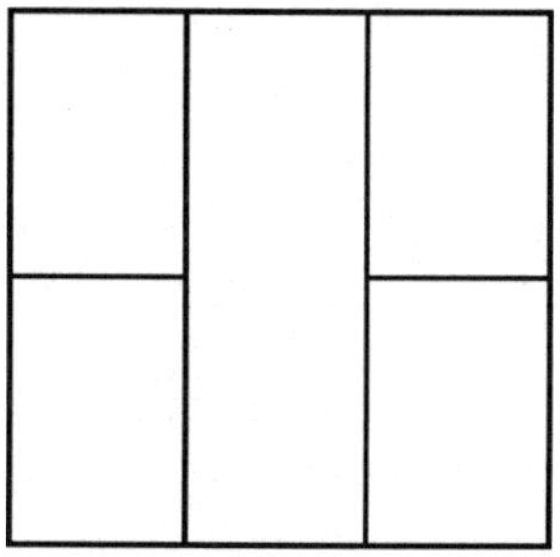

26. How can you cut a square layer cake that has been frosted on the top and all sides (but not the bottom) so that each piece will contain the same amount of frosting and the same volume of cake? There is more than one solution.

27. Out of construction paper, make small boxes that are $2 \times 1 \times 2.5$ units on a side. Fit as many of these boxes as possible in a larger box that is $4 \times 6 \times 5$ units. Describe your solution in writing. Is your solution unique?

CONSTRUCTIONS

28. Given a circle, use 3 line segments (curved, straight, open, or closed) to divide the circle into 8 sections.

EXTENSION: How can this be done so that all 8 sections have the same area?

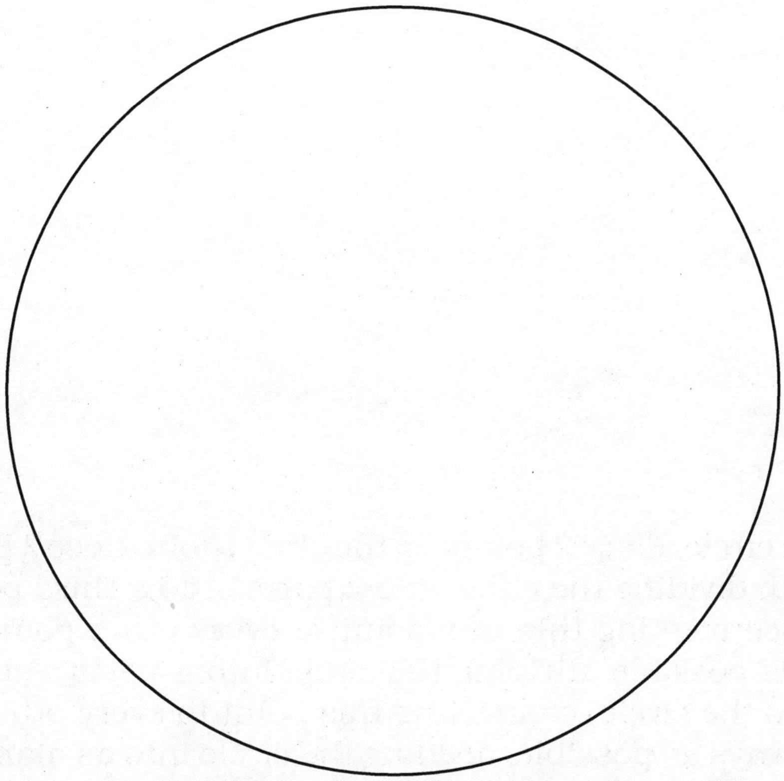

29. Draw a line segment on a piece of paper. Subdivide that segment into 3 segments. Using the three segments as side lengths, construct a triangle.

30. Draw a circle. Place 2 points on the circle. Join those 2 points with a chord, dividing the circle into 2 parts. Add a third point to the circle, connecting this new point to every other point as many ways as possible, dividing the circle into 4 parts. Add a fourth point to the circle, connecting this point to every other point as many ways as possible, dividing the circle into as many parts as possible. Continue adding points and forming the maximum number of regions as each new point is joined with the previous ones in as many ways as possible. Is there a pattern?

31. Create a figure like the polygon ABCDE below. Create a point F about which the figure can be rotated. Then create a line segment FG and predict what the rotated and reflected view of the figure will be.

EXTENSION: Would the end result be the same if the initial figure were reflected then rotated about a point? Do similar activities, increasing the complexity of the figure used initially.

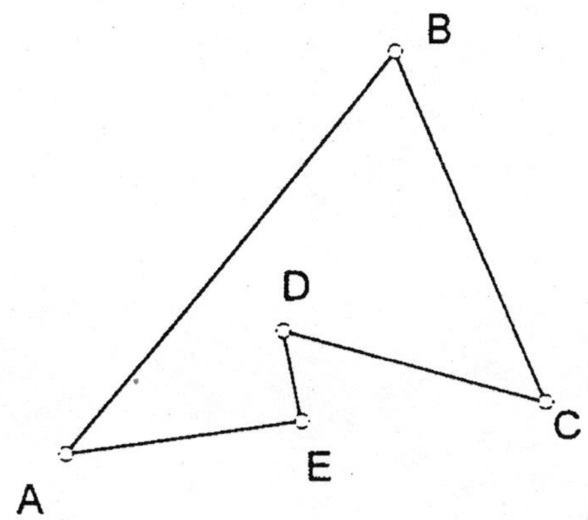

32. Construct a circle with center G. Mark points M, N, and H on the circle. Connect M and N with G, and M and N with H. Measure angle MHN and angle MGN. Move the points on the circle or change the radius of the circle. Write a conjecture describing what you have discovered.

NOTE: Software that permits moving points and changing radii of circles is particularly useful with this problem.

33. Inscribe congruent circles into a regular triangle, square, pentagon, hexagon, and decagon. What conclusion can be made about the respective areas of each of the inscribed figures?

34. Use a circle of a given radius to circumscribe a regular triangle, square, pentagon, hexagon and decagon. What conclusion can be made about the respective areas of each of the circumscribed figures?

35. A 9-point circle can be formed using any scalene triangle. The circle passes through the point at the base of each altitude, the midpoints of each side, and the midpoint of each segment joining a vertex to the point shared by the three altitudes. Do the construction.

NOTE: Geometry software is particularly useful with this activity.

36. Triangles can be shown to be congruent by a variety of methods. Which of the following properties can always be used to create a triangle congruent to a given triangle: SSS, SSA, AAA, ASA, SAS, SAA? Make a construction to demonstrate any property that will not always yield a triangle congruent to a given triangle.

NOTE: Geometry software is particularly useful with this activity because it permits the triangles to be altered.

37. Given a point, 6 equilateral triangles can be placed adjacent to each other so all space about that point is occupied. Since each angle of the equilateral triangle is 60°, the sum of all the angles about the point is 360°. Given any triangle, can it be arranged so that 6 identical triangles are placed so the sum of the angles about a point is 360°?

NOTE: Geometry software is particularly useful with this activity because it permits the triangles to be altered.

38. Create a large scalene triangle. In that triangle, construct the three medians. Write a conclusion about the medians. Measure the portion of each median that falls on either side of the point of concurrence, and write a conclusion about the ratio of the lengths.

NOTE: Geometry software is particularly useful with this activity because it permits the triangle and medians to be altered.

39. Create a large scalene triangle. In that triangle, construct the three angle bisectors. Write a conclusion about the angle bisectors.

NOTE: Geometry software is particularly useful with this activity because it permits the triangle and angle bisectors to be altered.

40. Create a large scalene triangle. In that triangle, construct the three altitudes. Write a conclusion about the altitudes.

NOTE: Geometry software is particularly useful with this activity because it permits the triangle and altitudes to be altered.

41. Create a large scalene triangle. In that triangle, construct the perpendicular bisector for each side. Write a conclusion about the perpendicular bisectors.

NOTE: Geometry software is particularly useful with this activity because it permits the triangle and perpendicular bisectors to be altered.

42. Create a large scalene triangle. In that triangle, construct the angle bisector for each angle, the altitude to each side, the perpendicular bisector for each side, and the median for each side, marking the point of concurrence for each set. Which one of the four points of concurrence is not always collinear with the other three? The line through the three collinear points is called the Euler Line, named after Leonard Euler (1707–1783), who was the first to prove the three points concurrent.

NOTE: Geometry software is particularly useful with this activity because it permits the triangle to be altered.

EXTENSION: Do the same construction with an isosceles triangle and an equilateral triangle. In each case, describe conclusions you reach.

43. Create a scalene triangle with exterior angles as shown. Measure the exterior angles. Write a conclusion about the sum of the measures of the exterior angles.

NOTE: Geometry software is particularly useful for this activity because it permits changing the triangle, and the calculator feature shows the measurements changing in real time.

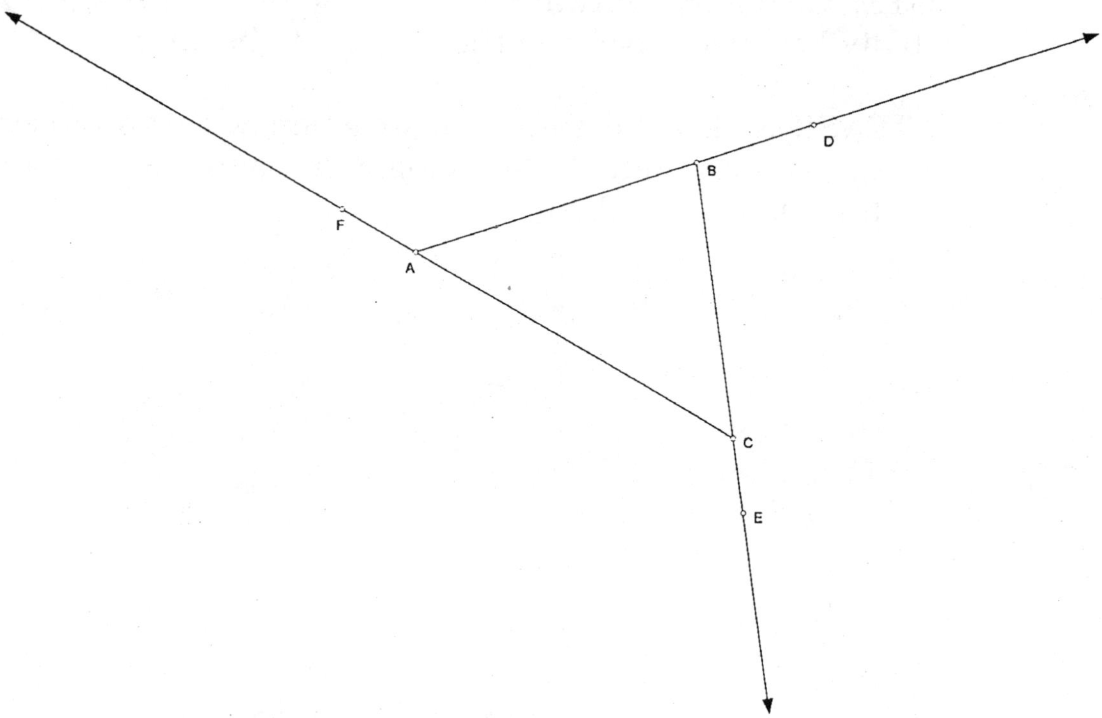

44. Create a quadrilateral with exterior angles as shown. Measure the exterior angles. Write a conclusion about the sum of the measures of the exterior angles.

NOTE: Geometry software is particularly useful for this activity because it permits changing the quadrilateral, and the calculator feature shows the measurements changing in real time.

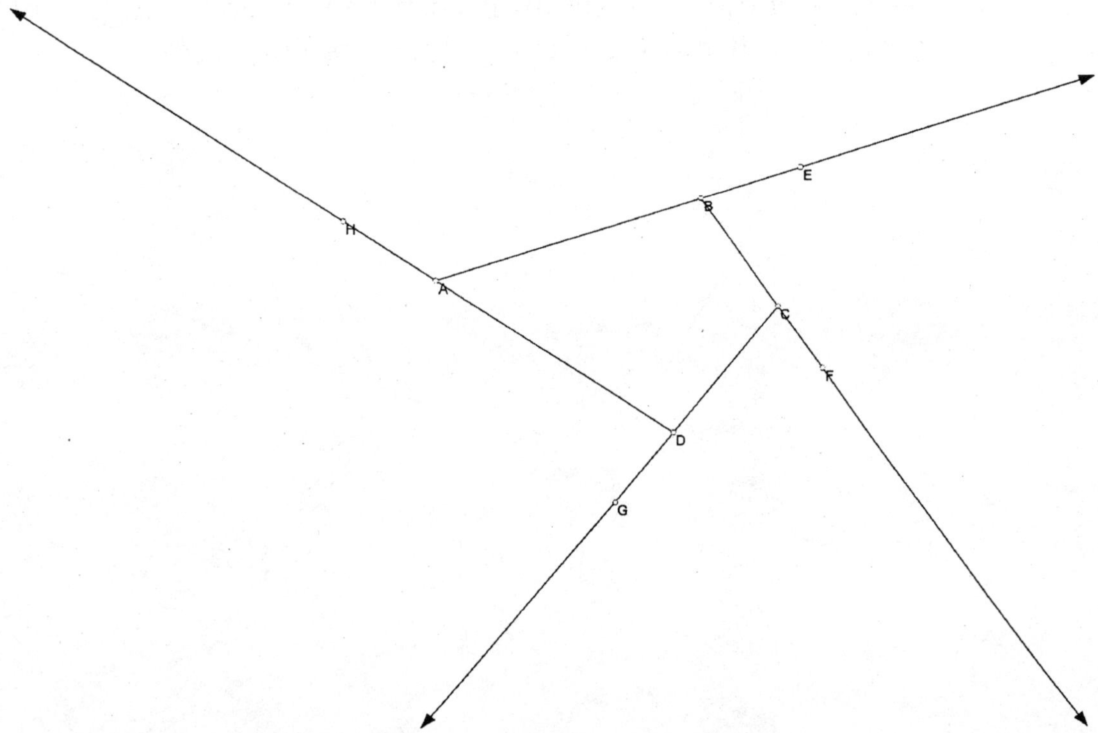

45. Create a pentagon with exterior angles as shown. Measure the exterior angles. Write a conclusion about the sum of the measures of the exterior angles.

NOTE: Geometry software is particularly useful for this activity because it permits changing the triangle, and the calculator feature shows the measurements changing in real time.

EXTENSION: Using this problem along with problems 43 and 44, write a conjecture dealing with the exterior angles of a polygon. Does it matter if the polygon is concave?

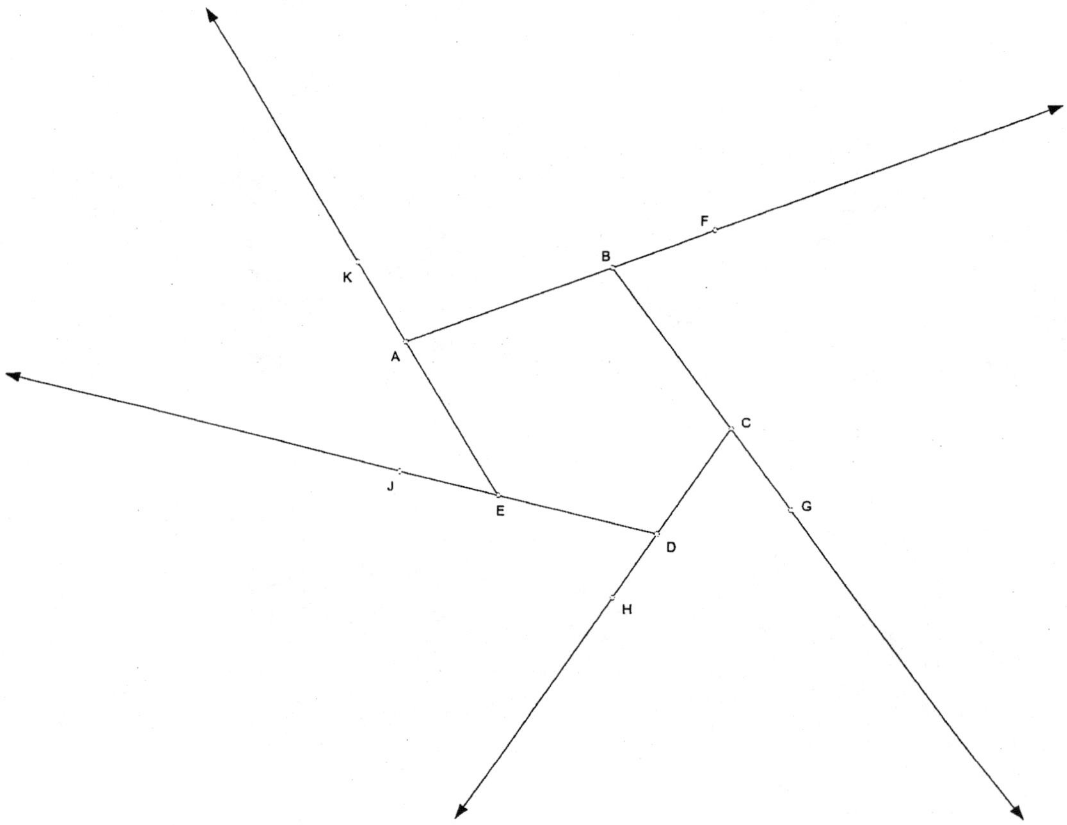

46. Cut a scalene triangle out of a piece of construction paper. Label the three vertices and tear them off the triangle. Place the three vertices together so they share a common point. After the first piece is placed, place each new piece next to an existing piece. Describe your result.

EXTENSION: Do this activity using a piece of software.

47. Cut a quadrilateral out of a piece of construction paper. Label the four vertices and tear them off the quadrilateral. Place the four vertices together so they share a common point. After the first piece is placed, place each new piece next to an existing piece. Describe your result.

EXTENSION: Do this activity using a piece of software.

48. Cut a pentagon out of a piece of construction paper. Label the five vertices and tear them off the pentagon. Place the five vertices together so they share a common point. After the first piece is placed, place each new piece next to an existing piece. Describe your result.

EXTENSION: Do this activity using a piece of software.

49. Cut a hexagon out of a piece of construction paper. Label the six vertices and tear them off the hexagon. Place the six vertices together so they share a common point. After the first piece is placed, place each new piece next to an existing piece. Describe your result.

EXTENSION: Do this activity using a piece of software.

50. Compare the results from problems 46 through 49. Describe the generalization. Are the results different for polygons that are concave?

51. Squares can be used to cover a plane with no overlaps and no gaps or spaces (this is called a pure tessellation of a plane). An example would be a piece of graph paper. Equilateral triangles can be used to tessellate a plane as well. Squares and equilateral triangles are regular polygons. Name another regular polygon that can be used to tessellate a plane. Other shapes, that are not regular, can also be used to tessellate a plane. A rectangle would be an example. Create an n-gon (a polygon with n sides) where $n > 4$ that will tessellate a plane, and sketch your result. If more than one shape is used, the shapes are said to tessellate the plane, but because more than one shape is used, it is not a pure tessellation. A flattened version of a soccer ball is an example. Use two or more shapes to tessellate a plane and sketch your result.

EXTENSION: Find examples of the tessellations of the Dutch artist M. C. Escher (1898–1972).

52. In 1796, the German mathematician Karl Friedrich Gauss (1777–1855) proved that by using just a straightedge and a compass, a regular n-gon (a polygon with n sides) can be constructed only when n is odd and either a prime Fermat number (a prime of the form $2^{2^n} + 1$) or the product of Fermat primes. Find at least 3 such n-gons and construct one n-gon.[1]

[1]Newman, James R. *The World of Mathematics*. New York: Simon and Schuster, 1956, 303–4.

VECTORS

53. Use an odd number of values, each of which represents the length of a vector. Start with the first vector and move horizontally to the right. Upon reaching the end of this first vector, turn 90° to the right and go the length of the second vector. Continue making a 90° right turn at the end of each vector. After a number of cycles through the set of vectors, the end of the last vector will be at the starting point of the activity. How many cycles does this take?

NOTE: A piece of software that allows repeated plotting of vectors can be very helpful with this problem.

54. Use an even number of values, each of which represents the length of a vector. Start with the first vector and move horizontally to the right. Upon reaching the end of this first vector, turn 90° to the right and go the length of the second vector. Continue making a 90° right turn at the end of each vector, except the last one. After a number of cycles through the set of vectors, the end of the last vector will be at the starting point of the activity. How many cycles does this take?

NOTE: A piece of software that allows repeated plotting of vectors can be very helpful with this problem.

EXTENSION: There are some exceptions to this generalization. Can you find one?

55. A tracker followed a bear 3 miles south, 3 miles west, and then 3 miles north. After the 9-mile trek, the tracker was back at the initial starting point. What color was the bear being tracked? Why?

56. Checker and chess boards are an 8×8 array of squares. In chess, the knight always follows an "L" pattern as it moves—two squares in one direction and one square perpendicular to the original path (or one square in one direction and two squares perpendicular to the original path). Assuming the knight is at its starting square (on a side, second square from corner), that its position is always at the center of a square, and that the squares are 2 inches on a side, what is the maximum distance away from the starting point the knight can be after 3 moves (assuming it does not get captured)? What is the minimum distance after 3 moves?

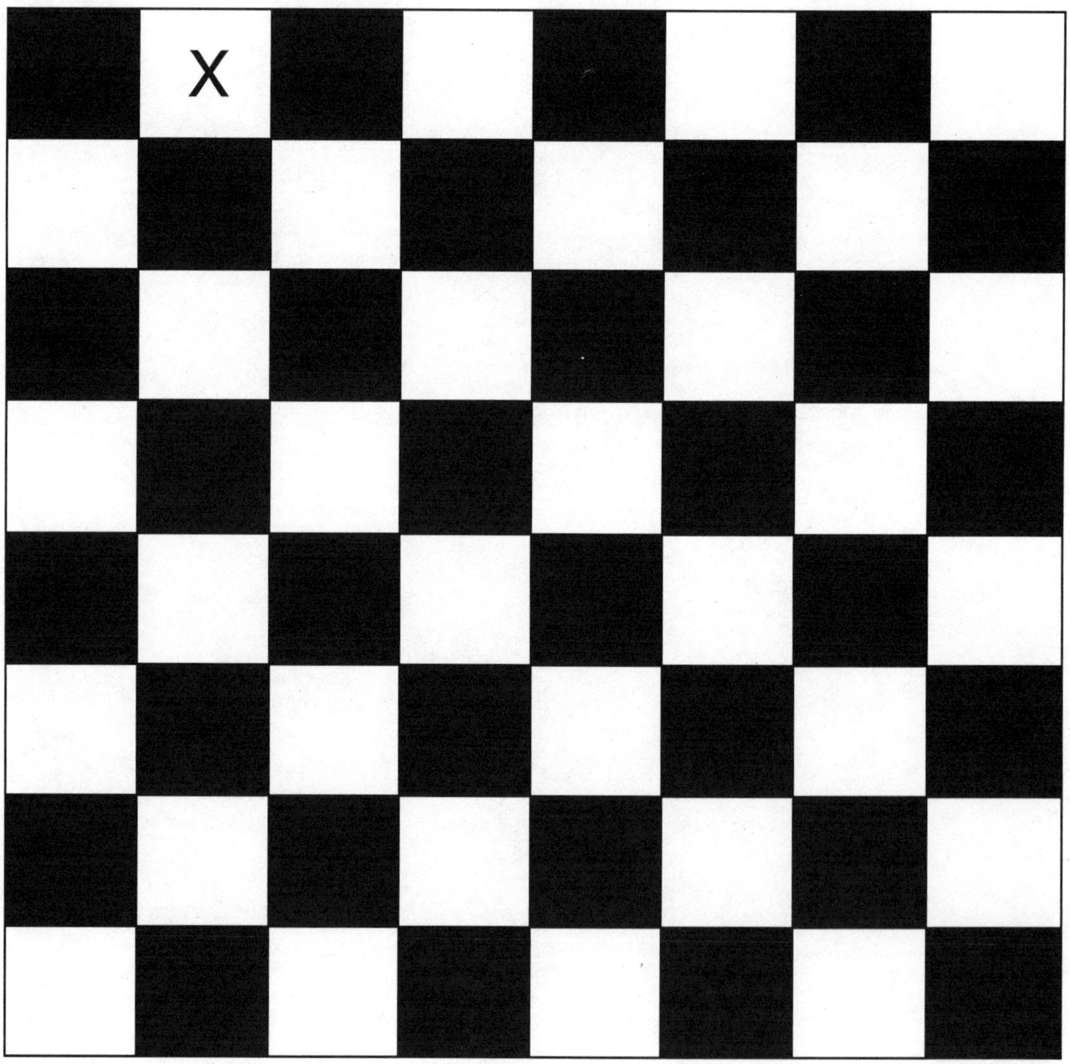

57. A mirror, a marble, and a tape measure can be used to measure the height of a tall object. Use the marble to assure the mirror is level. Stand a distance from the mirror, and then move so the top of the object can be seen in the mirror. Measure the distance from your eyes to the ground, the distance from the point beneath your eyes to the top of the object in the mirror, and the distance from the top of the object in the mirror to a point beneath the top of the object. Now, similar triangles have been established and the ratio of the distances will equal the ratio of the respective heights. Solve the proportion for the unknown height.[1]

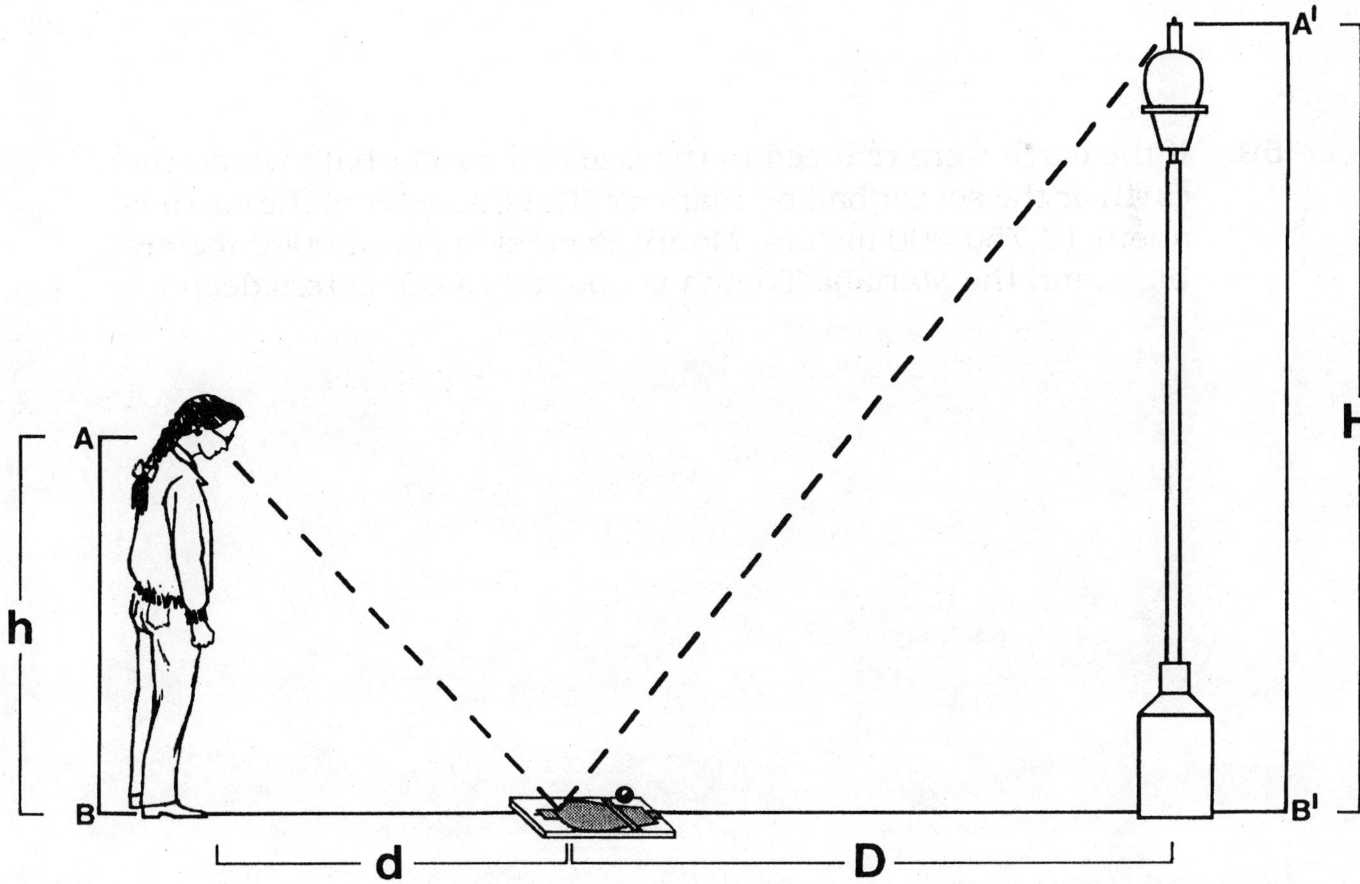

[1]Initial idea proposed by Kenneth Kidd, University of Florida

58. One person holds a ball representing the earth. A second person holds a ball representing the moon. How far apart do the two people need to stand to show the distance between the earth and moon?

59. If the earth were reduced to the size of a soccer ball, would the earth or the soccer ball be coarser? (The diameter of the earth is about 12,750,000 meters. Mount Everest is about 9000 meters high, and the Mariana Trench is about 11,000 meters deep.)

60. Suppose the earth is a sphere and its circumference at the equator is 25,000 miles. Imagine a ring that forms a band around the earth. It has a circumference of 25,000 miles plus 10 feet and the ring is concentric with the equator. What is the distance between the earth and the ring?

EXTENSION: Do the same problem with a 1-foot circle and an 11-foot circle.

61. Suppose a line segment (see a below) were divided into 2 parts (b and c) where the ratio of the entire segment length is to the longer segment (a to b) as the longer segment is to the shorter (b to c). This division satisfies the requirements for being a Golden Ratio. The Golden Ratio is $\frac{1+\sqrt{5}}{2}$ or 1.618034. The Fibonacci sequence can be used to provide approximations of the Golden Ratio. In a Fibonacci sequence (1, 1, 2, 3, 5, 8, 13, 21, 34, 55, 89, 144, 233, 377, 610,...), each new term is defined by adding the two preceding terms. Find decimal equivalents for adjacent number pairs in the Fibonacci sequence (for example, $\frac{13}{8}$, $\frac{21}{13}$, etc.). Describe a conclusion about adjacent pairs of Fibonacci numbers and the Golden Ratio.

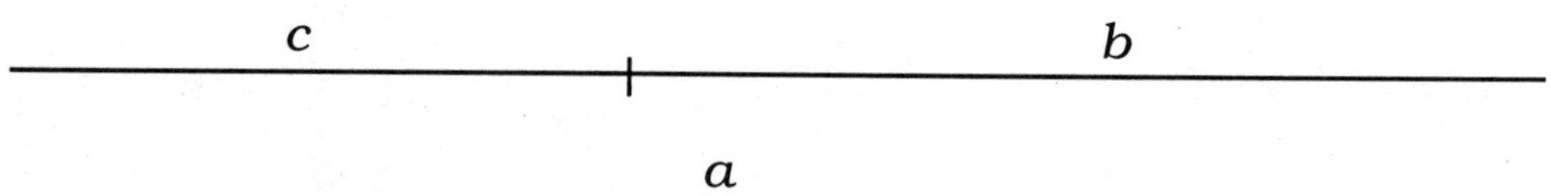

$\frac{a}{b} = \frac{b}{c}$

62. Imagine a canyon spanned by a bridge that is exactly one-mile long. The bridge has no expansion joints in it, so when it heats, it bows up. On one hot day, the bridge lengthened to one mile plus two feet. Approximately how high above its original unheated position is the midpoint of the bridge?

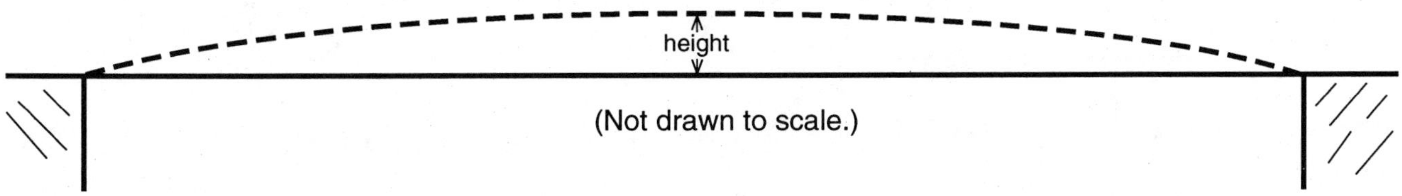

Solutions

1. Remove the two toothpicks that form either of the squares at the bend. One solution is shown below.

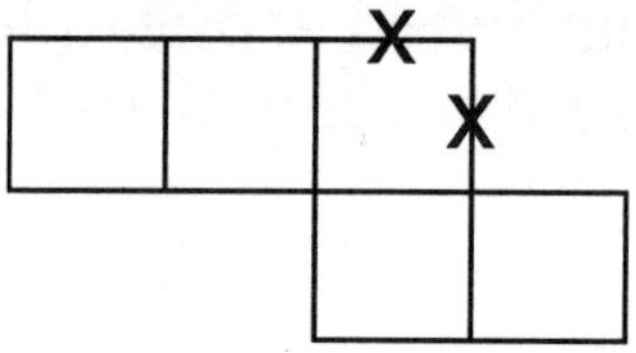

2.

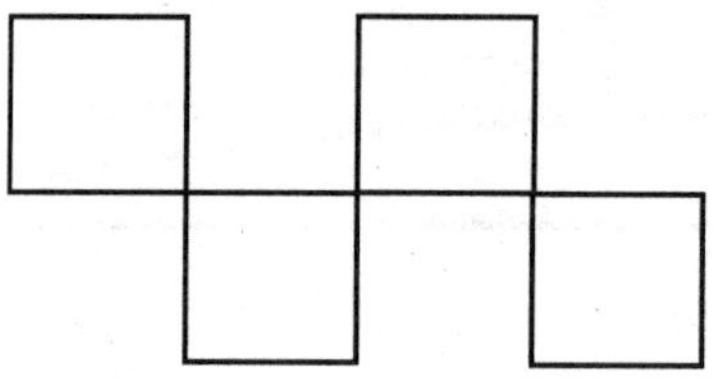

3. The top and bottom sides of the square shown in dotted lines are a whole toothpick-length long but are formed by two half-toothpicks.

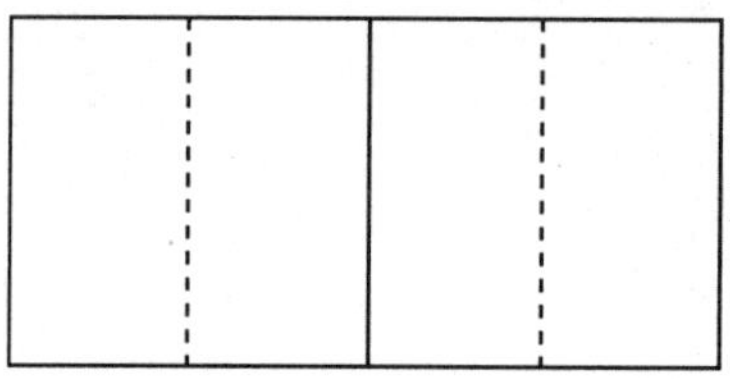

4. Overlap the 4 inner toothpicks slightly to create the figure below.

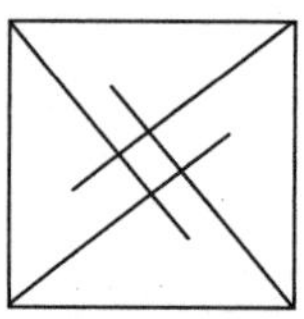

5. Six, five, or four toothpick lengths give 1 square; three toothpick lengths give 2 or 3 squares (a below); two toothpick lengths give 3 or 7 squares (b); and one toothpick length gives 6 or 7 (c). Note that the instructions did not say that the squares had to be congruent.

a.

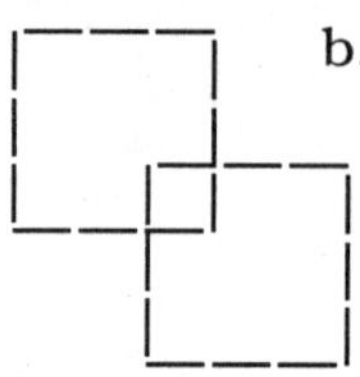

b.

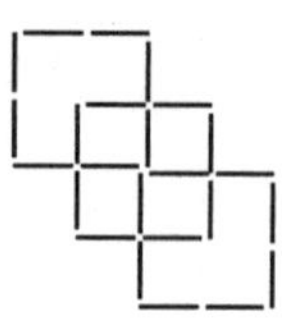

c.

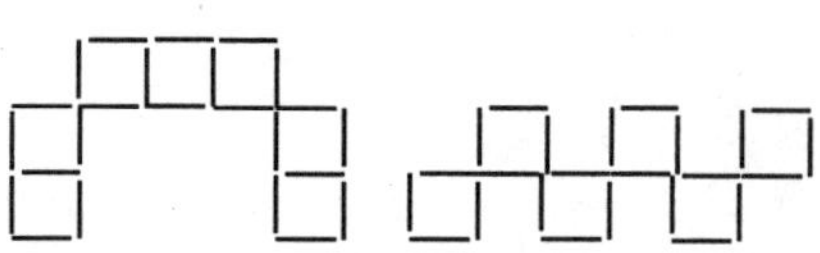

6. The four congruent triangles are four sides of an equilateral pyramid as shown below. The top of the drawing is the apex of the pyramid.

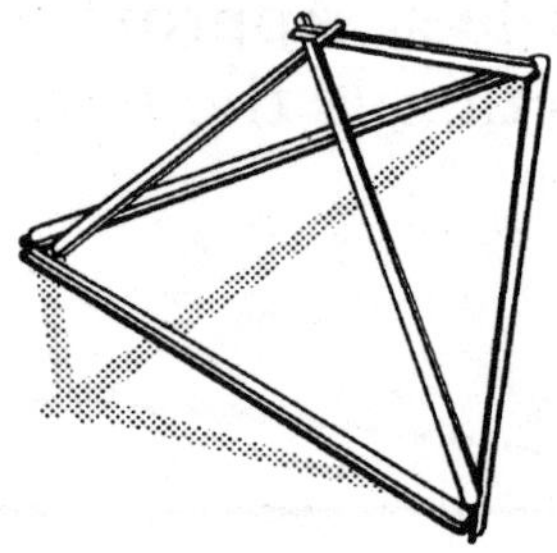

7.

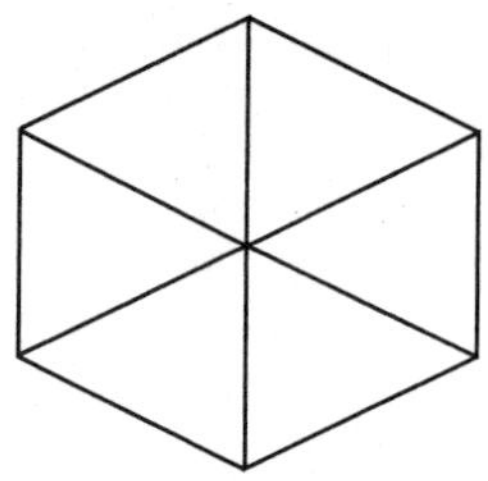

8.

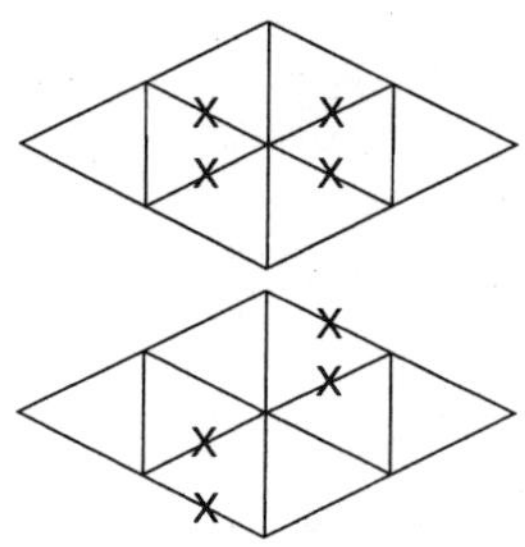

9. Accept any reasonable answers. Encourage students to use as much detail as possible, noting plane figures (circles, ellipses, triangles, rectangles, trapezoids, parallelograms, polygons, conic sections) as well as angles, conics, and solids.

10. The size of the holes does not matter; however, you may get some variation from the answers that follow depending on the initial setup. A circle gives an ellipse. A square gives an isosceles trapezoid. An equilateral triangle and an isosceles triangle both give isosceles triangles. A parallelogram gives a quadrilateral.

11. There are 204 squares: one 8 × 8 square, four 7×7, nine 6×6...sixty-four 1 × 1; 1 + 4 + 9 + 16 + 25 + 36 + 49 + 64 = 204.

12.

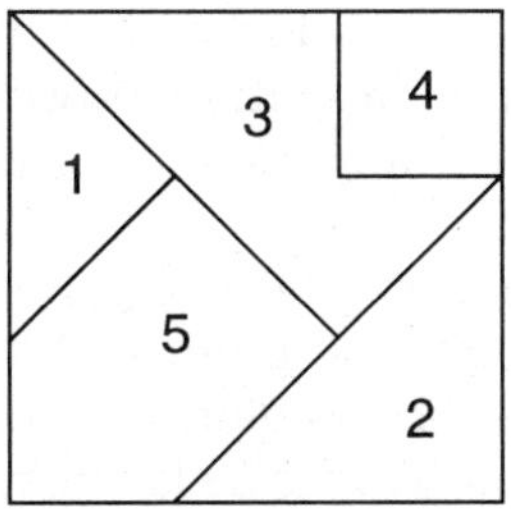

13.

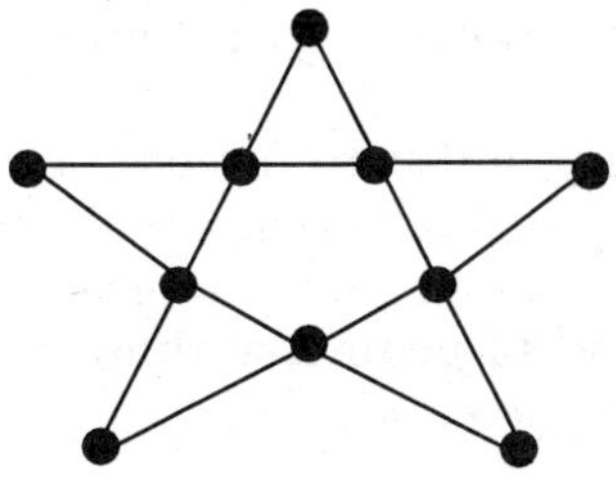

14. There is no triangle since (30 + 45) < 85.

15.

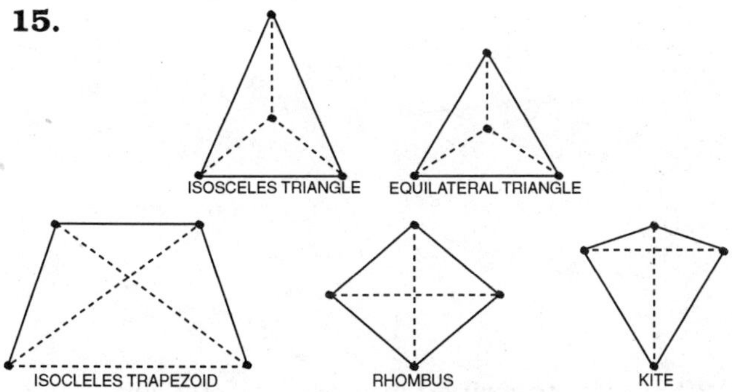

16.

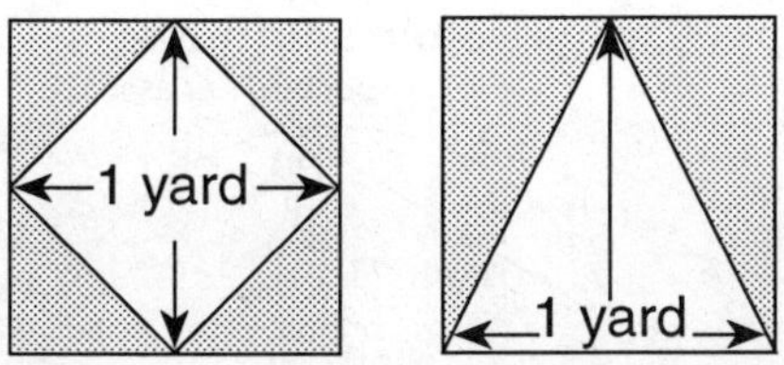

17. Responses may vary: What is the area? What is the perimeter? What is the maximum perimeter? What is the minimum perimeter? How many squares in all? Which shapes, when folded, form a box with no lid?

18. Each face of the MONTH cube holds two months, one written upside down. Each weekday is on one face of the DAY cube, and Saturday and Sunday share a face. The day cubes present some interesting problems. Assume both cubes must be used at all times. That means each must hold 0, 1, and 2. The fourth of a month, for example, would be represented as 04. That leaves six faces and seven digits. The problem can be solved by using the 6 as a 9, turning that particular face whichever way is necessary.

19. Starting 2 feet from the bottom right and top left corners of opposite long sides of the original rug, cut 1-foot high and 2-feet wide until the hole is reached from each side. Move the right-hand piece up and to the left to fit the pieces together.

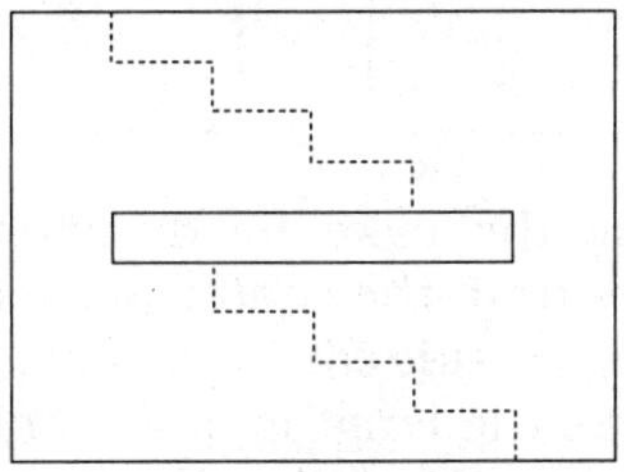

20. Manhole covers are round to prevent the lid from falling down the hole. An equilateral triangle will not work because the altitude is less than the length of a side and, if the altitude line were held parallel to the plane of the hole, the lid would fall through the hole when the lid was held sufficiently close to the side of the hole. Other figures with an odd number of sides will work. As the number of sides of a regular polygon increases, the figure will begin to react like a circle.

21. The circumference divided by the diameter should approximate π (3.141592654).

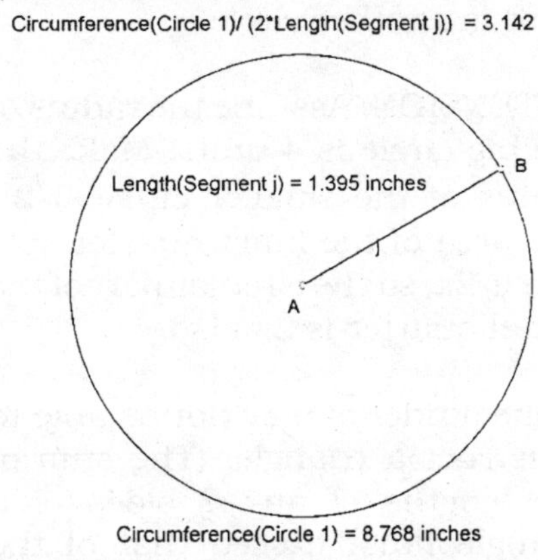

22. Length = πr, Height = r, Area = πr^2

23. $D = 6$ inches, area = 9π; twice that is 18π which is an 8.5-inch pizza

24. \$4.00—The area of the first bundle equals π; the area of the second bundle equals 4π ($A = \pi r^2$).

25. The side view could be divided evenly into quadrants, or it could be more unusual, as shown below.

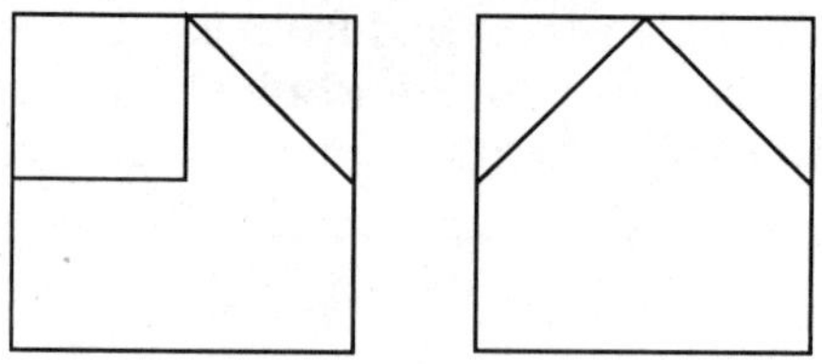

26. Divide the cake by drawing segments from the center to each corner. Each side can be subdivided as long as the lengths of the subdivisions are the same. Each triangle has the same base and height, so they all have the same area. Assuming the cake is flat, the volume will also be the same for each piece.

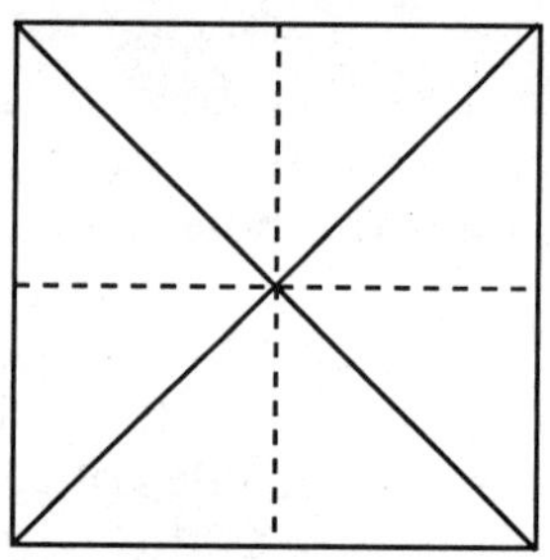

27. 24. The answer is unique.

28. Two of the segments are diameters set perpendicular to each other; the third is another circle.

EXTENSION: Assume the radius of the big circle is 4 units. Make the radius of the smaller circle $2\sqrt{2}$. The area of the inner quarter section is 2π, so the circular part of the initial quarter is 2π also.

29. Some students may not be able to construct a triangle. (The sum of the lengths of any 2 sides of a triangle must exceed that of the third side.) This exercise can be used to develop a probability discussion: How many students were able to construct a triangle?

30. Two points give 2 regions; 3 points, 4 regions; 4 points, 8 regions; 5 points, 16 regions; 6 points, 31 regions. This is a classic example of a pattern breaking down. The moral is to be sure to gather enough data.

31. Using software forces students to become conversant with computers and requires a greater command of geometry than doing the exercise concretely.

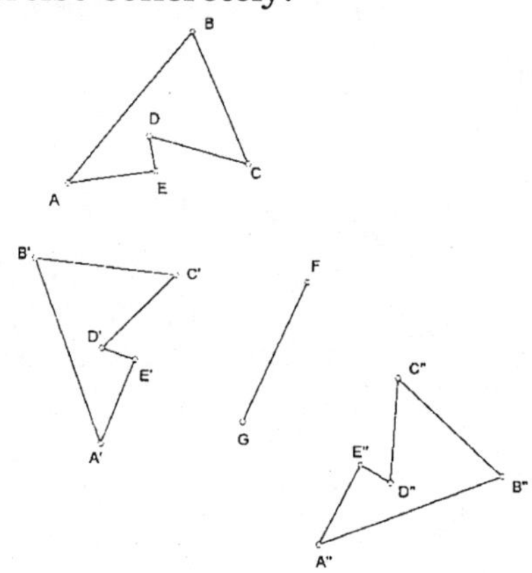

32. The inscribed angle is half the central angle when they cut the same arc. Two possible solutions:

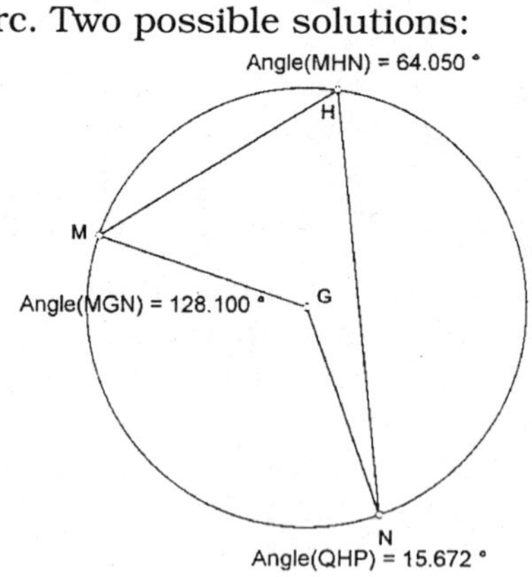

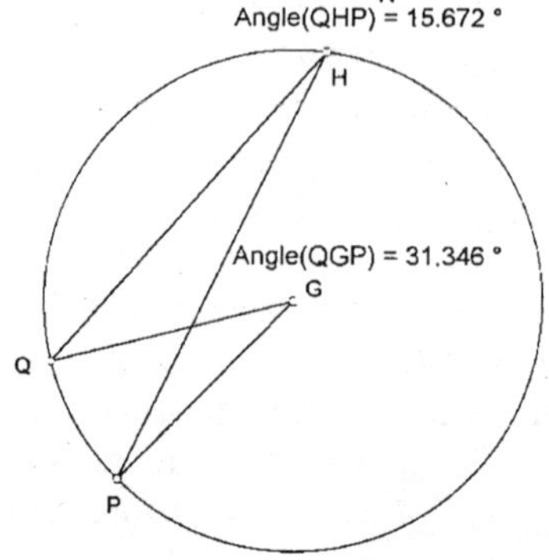

33. As the number of sides increases, the area of the regular polygon increases, approaching the area of the circle.

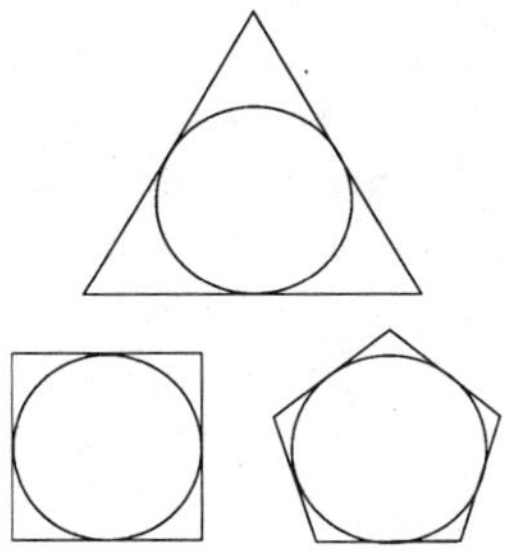

34. As the number of sides increases, the area of the regular polygon increases, approaching the area of the circle.

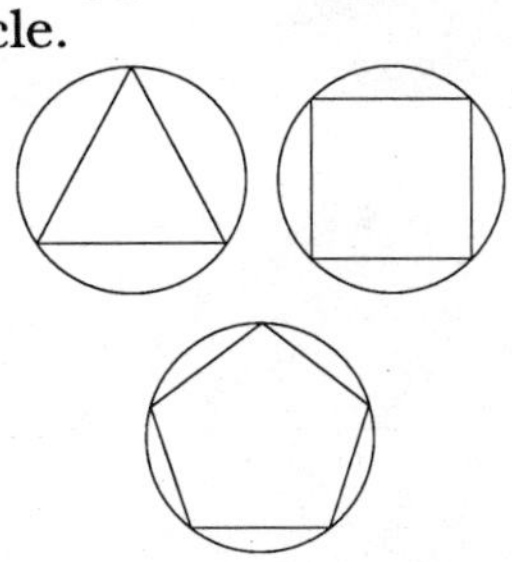

35. In the solution below, D, E, and F are midpoints of sides; G, H, and J are the bases of altitudes; K is the intersection point of the altitudes; L, M, and N are midpoints of segments AK, BK, and CK; P is the center of the circle, made by the connecting the midpoints D, E, and F with their respective opposite midpoints (N, L, and M) on segments CK, AK, and BK. Line segments PD, PE, and PF are radii of the circle that passes through the nine points: D, J, M, G, E N, F, H, and L.

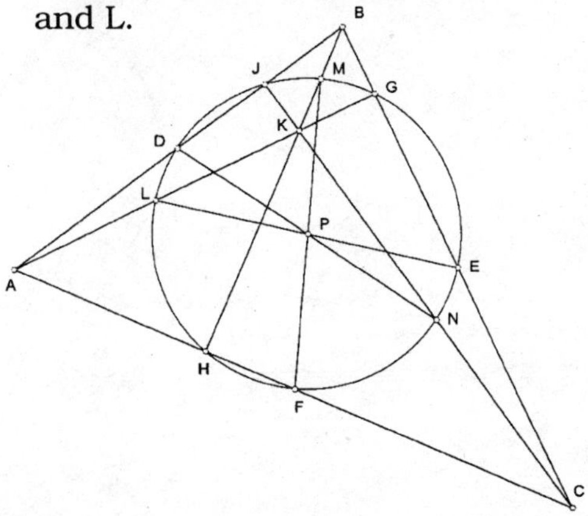

36. As shown below, AAA and SSA do not always give congruent triangles.

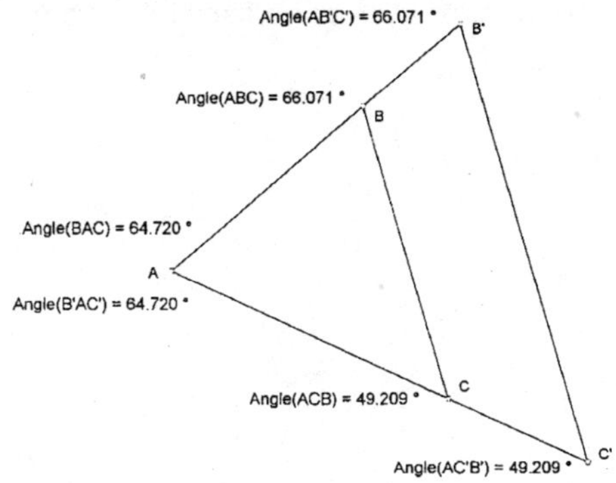

Angle BAC and side AB are constant. Side BC' is the same size as side BC. Triangles ABC and ABC' share SSA but are not congruent.

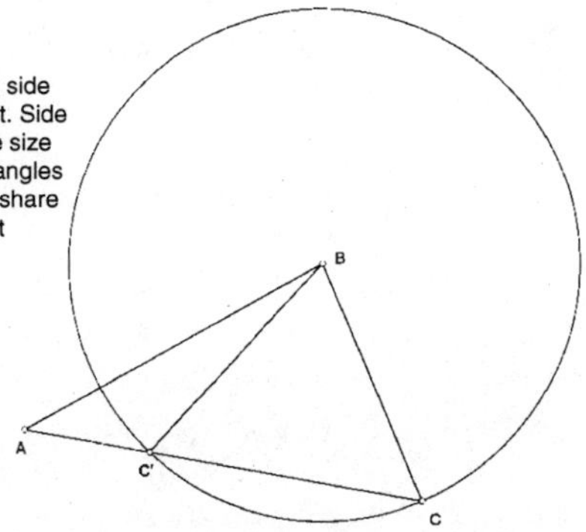

37. Yes. The first diagram below shows 6 copies of an equilateral triangle around point A. The second diagram shows 6 copies of scalene triangle ABC around point D.

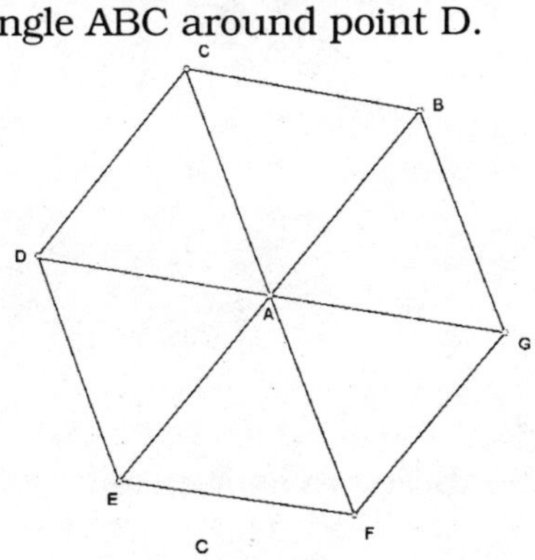

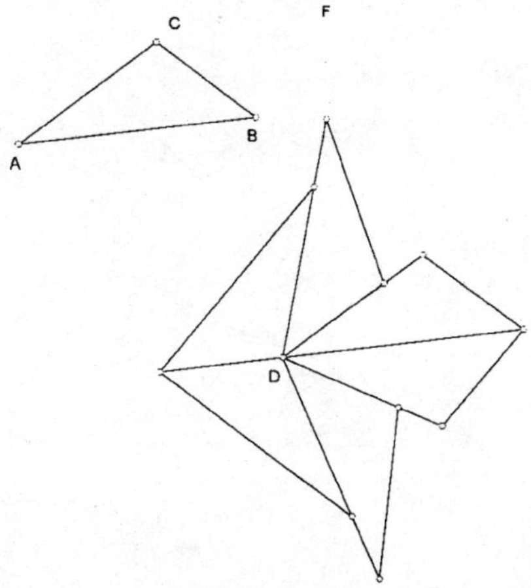

38. The three medians are always concurrent, and the point of concurrence divides each median in a ratio of 2:1.

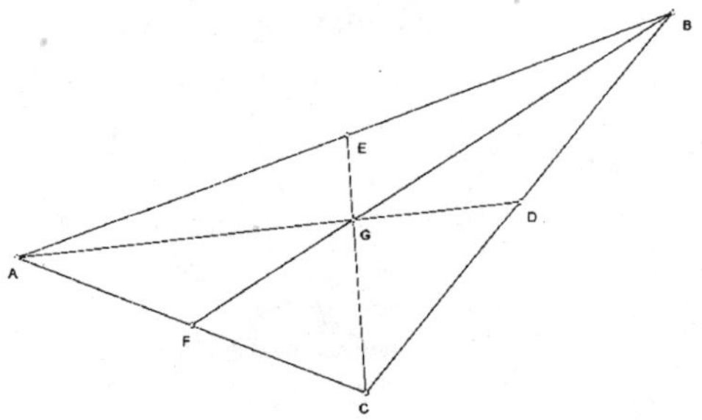

39. The three are concurrent.

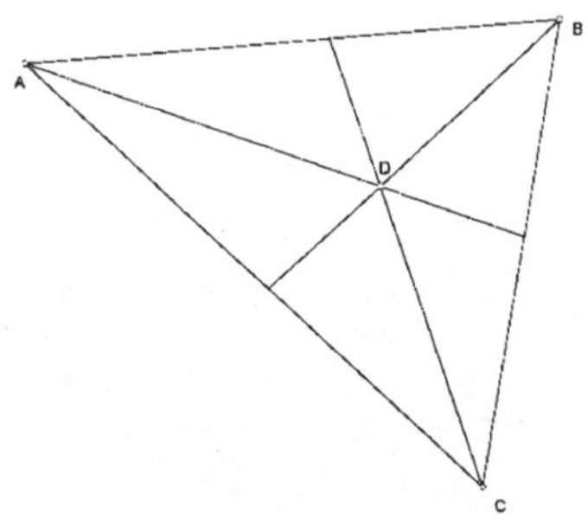

40. The three are concurrent.

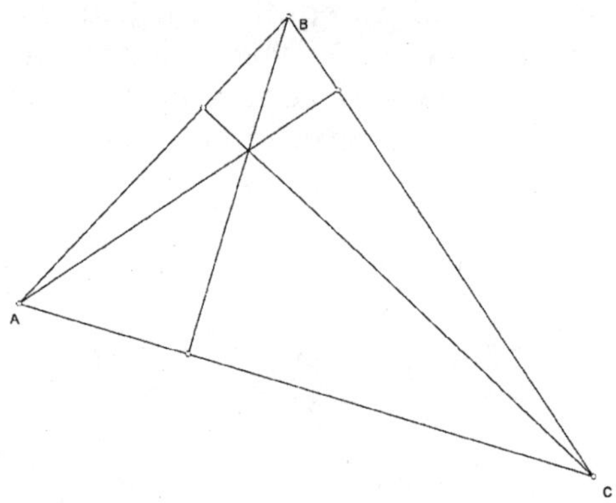

41. The three are concurrent.

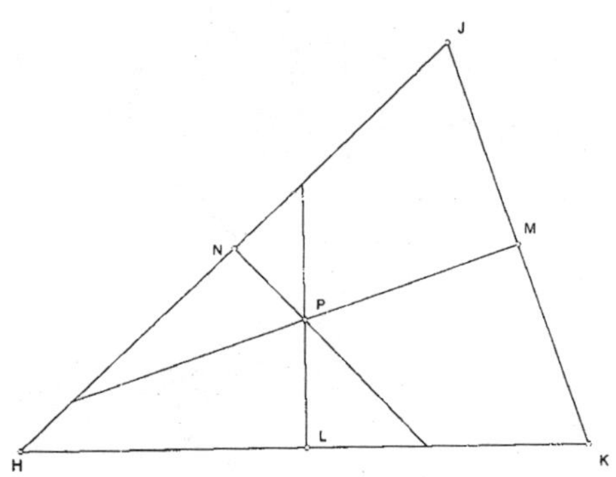

42. In an equilateral triangle (first diagram below), the medians, midlines, altitudes, and angle bisectors all are concurrent. In an isosceles triangle (second diagram below), the concurrence of the medians, midlines, altitudes, and angle bisectors form an Euler Line. In scalene triangles (third diagram below), the angle bisector (Point Q below) will not always be on the Euler Line.

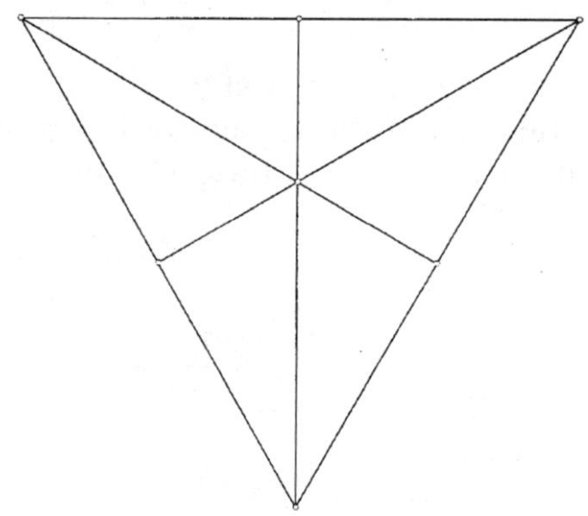

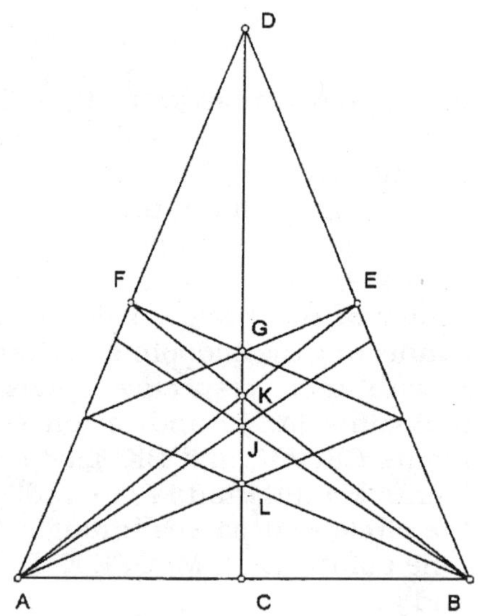

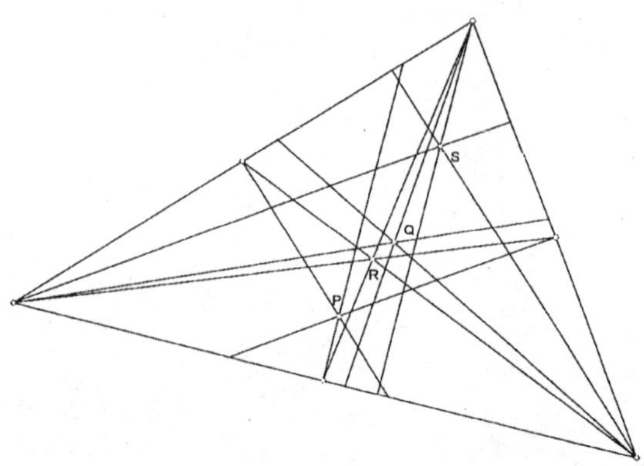

43. In the diagram below, angle FAB = 132°, angle DBC = 100°, and angle ECA = 128°. The sum of the exterior angles is 360°.

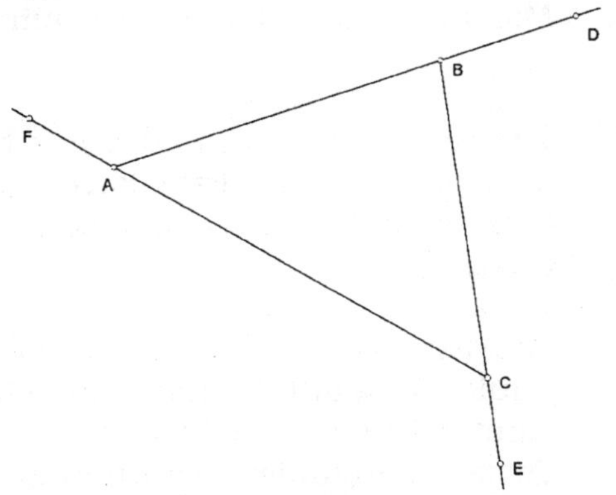

44. In the diagram below, angle HAB = 131°, angle EBC = 71°, angle FCD = 76°, and angle GDA = 82°. The sum of the exterior angles is 360°.

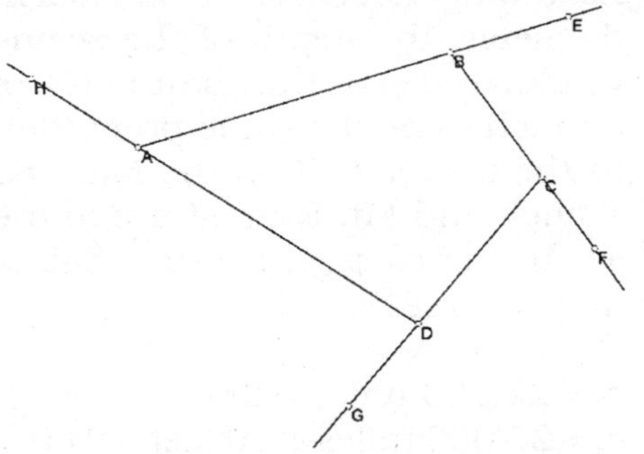

45. In the diagram below, angle KAB = 131°, angle FBC = 71°, angle GCD = 66°, angle HDE = 65°, and angle JEA = 27°. The sum of the exterior angles is 360°.

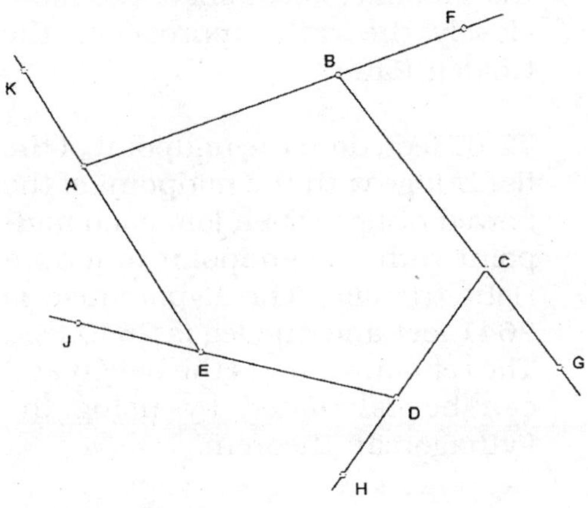

46. The three vertices form a straight (180°) angle.

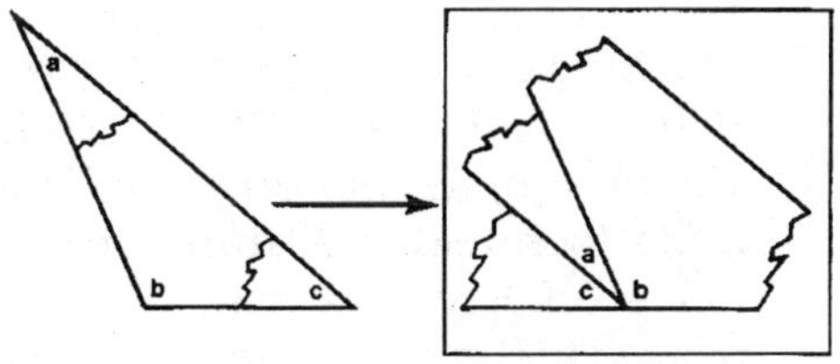

47. The four vertices form a 360° angle.

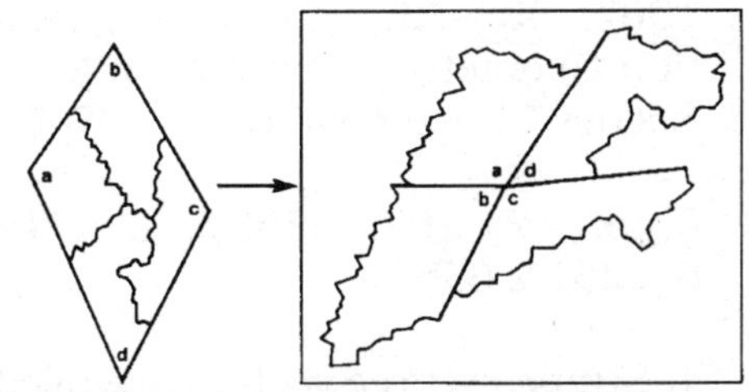

48. The five vertices form a 540° angle.

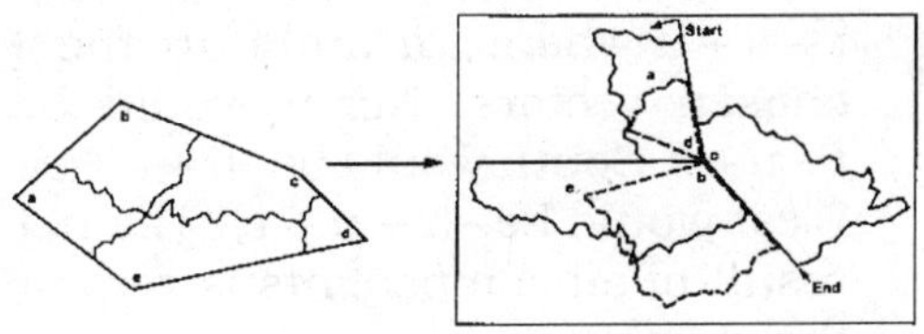

49. The six vertices form a 720° angle.

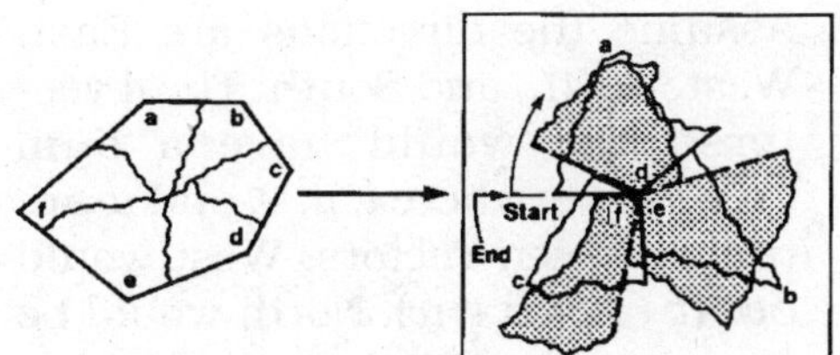

50. Whether the polygon is concave or convex, the angle formed = $(n-2) \times 180°$, where n = the number of sides in the polygon.

51. A regular hexagon tessellates a plane. Other responses will vary.

EXTENSION: The artist M. C. ESCHER was born in the Netherlands on June 17, 1898. He died March 27, 1972. He explored geometry in his artwork, including many tessellations. Sources include *Non-*

Euclidean Tessellations and their Groups by Magnus, *M. C. Escher: 29 Master Prints* by Escher, *The World of M. C. Escher* by Escher, *The Magic Mirror of M. C. Escher* by Bruno Ernst, and *M. C. Escher: Visions of Symmetry* by Doris Schattschneider. A number of articles on Escher's work have appeared in the *Mathematics Teacher*: "How to Draw Tessellations of the Escher Type," April 1974, 307–10; "Mathematics and Art," October 1967, 568–72; and "Transformation Geometry & Artwork of M. C. Escher," December 1976, 647–52.

52. 3, 5, 17, 257, 65537, 3 × 17, 5 × 257 × 65537

53. 4. Suppose three vectors are used. Assume the directions are East, West, North, and South. The 3 vectors East would have a sum $t + u + v$ where t, u, and v are the 3 chosen vectors. North would be $t + u + v$. South would be $-(t + u + v)$. West would be $-(t + u + v)$. The net result of all components is zero or the starting point.

54. 4. Suppose four vectors are used. Assume the directions are East, West, North, and South. The 4 vectors East would have a sum $t + u + v + w$ where t, u, v, and w are the 4 chosen vectors. West would be $-(t + u + v + w)$. North would be $t + u + v + w$. South would be $-(t + u + v + w)$. The net result of all components is zero or the starting point.

If two vectors are used and if a 90° right turn is made after the second vector, the starting point is achieved after two executions.

If the vector lengths are the same, the starting point is achieved after 2 cycles for 2 vectors, 1 cycle for 4 vectors, 2 cycles for 6 vectors.

55. The only place this can happen is at the north pole, so the bear must be white (polar bear).

56. Maximum 13.4164 inches; minimum 2 inches

57. The values for h, d, and D can be measured. H can be calculated from these values: $H/D = h/d$, so $H = Dh/d$.

58. The distance between the two balls would be about 10 times the circumference of the earth ball or 35 times the circumference of the moon ball. (Note: To be proportional, the diameter of the earth ball would need to be 4 times that of the moon ball.)

59. Assuming the soccer ball is 1 foot in diameter, the depth of the seams would need to be less than 1/100 of an inch to be the same proportion to the soccer ball as the Mariana Trench and Mt. Everest are to the earth; therefore, the soccer ball is more coarse.

60. $c_e = 25{,}000$ miles $= 2\pi r$
$c_b = 25{,}000$ miles + 10 feet $= 2\pi(r + x) = 2\pi r + 2\pi x$
But 25,000 miles $= 2\pi r$. Therefore, 10 feet $= 2\pi x$. Divide both sides by 2π, and $x = 1.59$ feet.

61. The larger the pairs of numbers in the Fibonacci sequence, the more closely the ratio approaches the Golden Ratio.

62. 72.67 feet. Join the midpoint of the flat bridge with the midpoint of the bowed bridge. Then join each midpoint with one endpoint to form a right triangle. The hypotenuse is 2641 feet and one leg is 2640 feet. The remaining leg is the height and can be calculated by using the Pythagorean theorem.